AMAZING SPACE

THE MOON

BY MATT LILLEY

CREATIVE EDUCATION • CREATIVE PAPERBACKS

Published by Creative Education and Creative Paperbacks
P.O. Box 227, Mankato, Minnesota 56002
Creative Education and Creative Paperbacks
are imprints of The Creative Company
www.thecreativecompany.us

Design by The Design Lab
Art direction by Graham Morgan
Edited by Joe Tischler and Ana Brauer

Photographs by Getty/Grant Faint, 21, NASA, 14, NurPhoto, 5, Rolls Press/Popperfoto, 17; NASA/Unknown, 2, 9, 10, 13, 15, 16, Ben Smegelsky, 22–23, NOAA, 6; Pexels/Axel Sandoval, 4; Unsplash/ Michael, 19; Wikimedia Commons/ Justus Sustermans, 12, NASA/Neil Armstrong, 18, NASA's Scientific Visualization Studio - Ernie Wright, Noah Petro, David Ladd, 1, NASA's Scientific Visualization Studio - Ernie Wright, Noah Petro, David Ladd, 1, Paramanu Sarkar, cover

Library of Congress Cataloging-in-Publication Data
Names: Lilley, Matt, author.
Title: The moon / Matt Lilley.
Other titles: Amazing space (Creative Education, Inc. (Mankato, Minn.))
Description: Mankato, Minnesota : Creative Education and Creative Paperbacks, [2025] | Series: Amazing space | In cludes bibliographical references and index. | Audience: Ages 6–9 | Audience: Grades 2–3 | Summary: "Learn about Earth's moon and the discoveries astronauts have made there. This astronomy title for elementary-aged kids includes captions, on-page definitions, a space science feature, additional STEM resources, and an index"—Provided by publisher.
Identifiers: LCCN 2024018577 (print) | LCCN 2024018578 (ebook) | ISBN 9798889892373 (library binding) | ISBN 9781682776032 (paperback) | ISBN 9798889893486 (ebook) Subjects: LCSH: Moon—Juvenile literature.
Classification: LCC QB582 .L55 2025 (print) | LCC QB582 (ebook) | DDC 523.3—dc23/eng/20240509
LC record available at https://lccn.loc.gov/2024018577
LC ebook record available at https://lccn.loc.gov/2024018578

Printed in China

Table of Contents

The Moon is Earth's only natural **satellite.**

The Moon is the brightest object in the night sky. It reflects sunlight. We see the part of the Moon that the Sun is shining on. The different shapes we see are called phases.

satellite something that moves around a larger object

The Moon is about 238,900 miles (384,472 kilometers) from Earth. If you could drive there, it would take about six months. The same side of the Moon always faces the Earth. No one ever saw the far side until people took pictures from space.

Luna is the Latin word for Moon.

The Moon is rocky. It is mostly made of rocks similar to those on Earth. Scientists think the Moon used to be part of Earth. The Moon might have been formed when a planet the size of Mars crashed into Earth.

The oldest rock found on the Moon is over 4.4 billion years old.

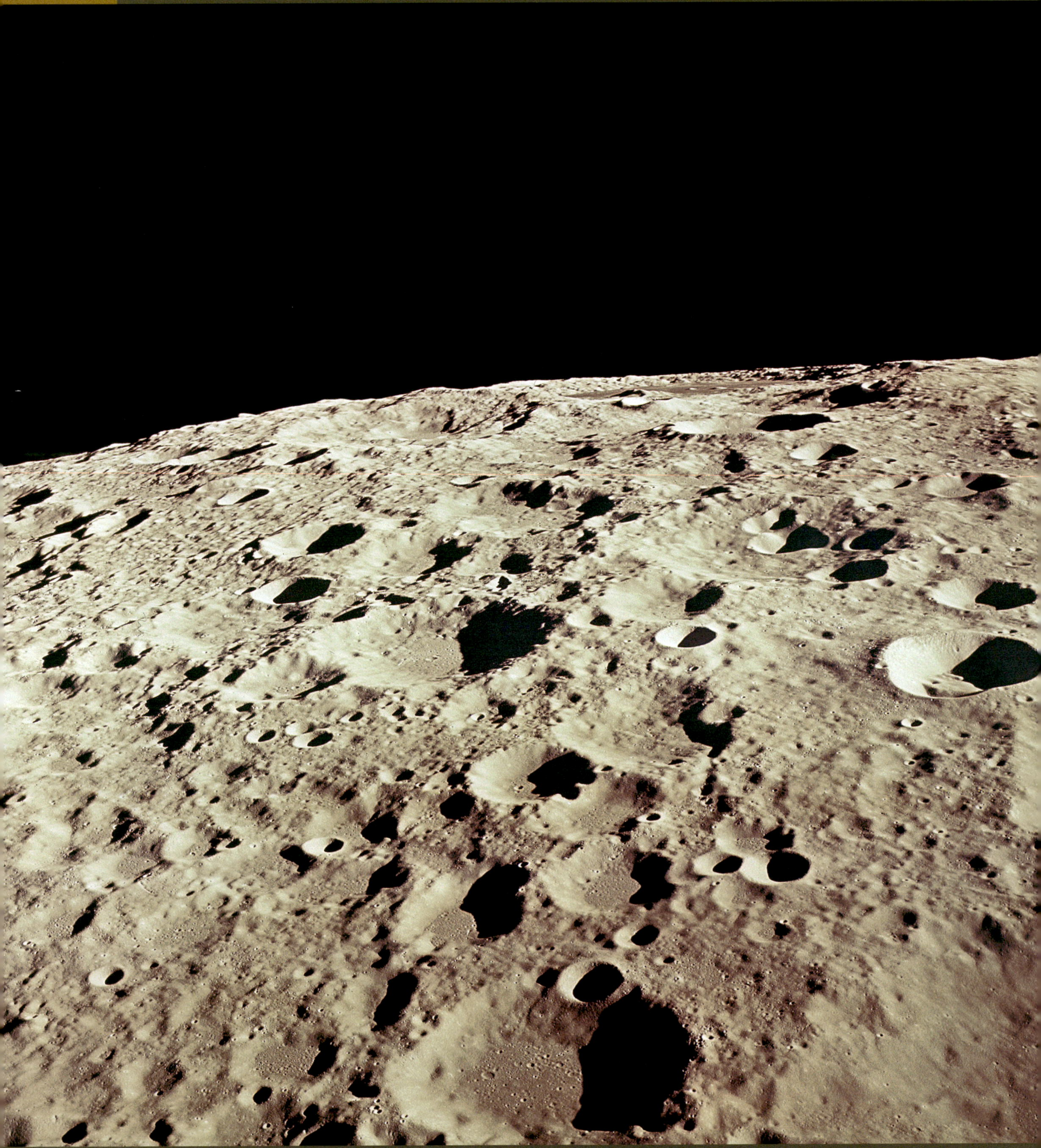

Craters are caused by meteorites or comets hitting the Moon's surface.

The Moon is covered in **craters**. It has almost no air or water. Because of this, the Moon gets very hot in the day and cold at night. There is not much gravity. Something that weighs 60 pounds on Earth (27 kilograms) would weigh 10 lbs (4.5 kg) on the Moon.

crater a bowl-shaped empty space in the ground

The Moon has many volcanoes. They are no longer active.

In 1609, Galileo Galilei made a telescope. He pointed it at the Moon. At the time, people thought the Moon was flat. Through the telescope, Galilei saw mountains. He saw valleys. He realized the Moon was similar to Earth.

valley a low area between hills or mountains

The footprints from the Apollo 11 astronauts are still on the Moon.

In 1969, the first astronauts landed on the Moon. They took pictures. They did **research**. The astronauts brought rocks back to Earth. Scientists are still studying those rocks.

research to study something

Six space missions landed men on the Moon. They drove rovers on the Moon for three of them.

The first astronauts were jet pilots. Most astronauts today are scientists or engineers. They visit space stations and do research. They also repair anything that might be broken on the station. Future astronauts will go back to the Moon.

While the Soviets never reached the moon, they were the first to put humans in space.

Like Earth, the Moon has north and south poles. At the poles, the Sun stays low. Some craters at the poles are always in shadow. They stay cold and dark. Those craters might contain ice. Future astronauts could use that water.

It takes about 27 Earth days for the Moon to orbit Earth.

Without the Moon, Earth would be very different. The Moon helps keep Earth steady as it rotates. Without the Moon, the planet would wobble more. If Earth wobbled too much, our seasons would change.

The Moon is slowly drifting away from Earth.

Space Spotlight: Artemis Moon Missions

The last human walked on the Moon in 1972. NASA's new Artemis program will bring people back to the Moon. Artemis II plans to fly around the Moon in 2025. In 2026, Artemis III plans to land on the Moon. The astronauts will land near the Moon's south pole. Later missions will build a Moon base. Astronauts will live there. From the Moon, they might someday go to Mars.

NASA
esa

Read More

Hirsch, Rebecca E. *Mysteries of the Moon*. Minneapolis, Minn. Lerner Publications, 2020.

Huddleston, Emma. *Explore the Moon*. Minneapolis, Minn. ABDO Publishing, 2021.

Schuh, Mari. *The Moon: Earth's Satellite*. Minneapolis, Minn. Jump!, 2023.

Websites

NASA Science Space Place: Moon
https://spaceplace.nasa.gov/search/moon/
Learn more about the Moon and play games.

Kiddle: Moon Facts for Kids
https://kids.kiddle.co/Moon
Learn more fun facts about the Moon.

Index